MATH
WORKBOOK

ADDITION

1. $11 + 84$ …………	2. $31 + 63$ …………	3. $27 + 27$ …………	4. $8 + 75$ …………
5. $64 + 32$ …………	6. $57 + 10$ …………	7. $44 + 69$ …………	8. $28 + 83$ …………
9. $44 + 38$ …………	10. $42 + 95$ …………	11. $45 + 48$ …………	12. $61 + 99$ …………
13. $28 + 34$ …………	14. $7 + 25$ …………	15. $29 + 98$ …………	16. $6 + 66$ …………
17. $47 + 15$ …………	18. $47 + 41$ …………	19. $4 + 88$ …………	20. $10 + 93$ …………
21. $59 + 46$ …………	22. $20 + 68$ …………	23. $58 + 87$ …………	24. $73 + 47$ …………

1. 23 + 54 = ...	2. 36 + 13 = ...	3. 24 + 97 = ...	4. 42 + 47 = ...
5. 54 + 69 = ...	6. 43 + 51 = ...	7. 92 + 41 = ...	8. 69 + 24 = ...
9. 8 + 81 = ...	10. 13 + 74 = ...	11. 56 + 18 = ...	12. 81 + 71 = ...
13. 96 + 58 = ...	14. 45 + 67 = ...	15. 33 + 42 = ...	16. 23 + 83 = ...
17. 5 + 54 = ...	18. 48 + 50 = ...	19. 96 + 96 = ...	20. 44 + 33 = ...
21. 90 + 91 = ...	22. 47 + 52 = ...	23. 49 + 99 = ...	24. 75 + 86 = ...

1. 139 + 143 = ...

2. 99 + 94 = ...

3. 129 + 141 = ...

4. 159 + 160 = ...

5. 199 + 107 = ...

6. 106 + 66 = ...

7. 92 + 99 = ...

8. 182 + 166 = ...

9. 141 + 146 = ...

10. 160 + 95 = ...

11. 73 + 142 = ...

12. 148 + 68 = ...

13. 90 + 111 = ...

14. 184 + 147 = ...

15. 166 + 192 = ...

16. 89 + 72 = ...

17. 110 + 93 = ...

18. 143 + 186 = ...

19. 62 + 131 = ...

20. 155 + 133 = ...

21. 87 + 75 = ...

22. 93 + 98 = ...

23. 111 + 173 = ...

24. 143 + 191 = ...

1. 190 + 77	2. 58 + 199	3. 99 + 100	4. 157 + 117
5. 168 + 193	6. 76 + 174	7. 191 + 145	8. 159 + 149
9. 189 + 146	10. 190 + 182	11. 127 + 137	12. 175 + 75
13. 81 + 87	14. 80 + 77	15. 66 + 144	16. 85 + 125
17. 88 + 191	18. 154 + 86	19. 153 + 106	20. 148 + 146
21. 123 + 137	22. 100 + 164	23. 186 + 70	24. 65 + 123

1. 163 + 120 = ...	2. 155 + 171 = ...	3. 162 + 111 = ...	4. 130 + 104 = ...
5. 193 + 125 = ...	6. 107 + 200 = ...	7. 172 + 108 = ...	8. 105 + 119 = ...
9. 161 + 142 = ...	10. 142 + 104 = ...	11. 187 + 143 = ...	12. 108 + 148 = ...
13. 149 + 144 = ...	14. 179 + 199 = ...	15. 163 + 182 = ...	16. 113 + 104 = ...
17. 159 + 152 = ...	18. 141 + 162 = ...	19. 163 + 166 = ...	20. 195 + 157 = ...
21. 145 + 179 = ...	22. 199 + 141 = ...	23. 194 + 105 = ...	24. 183 + 122 = ...

1.	2.	3.	4.
147 + 185	159 + 107	144 + 167	155 + 144
....	....	....	....

5.	6.	7.	8.
162 + 156	144 + 124	100 + 162	100 + 177
....	....	....	....

9.	10.	11.	12.
113 + 121	150 + 156	143 + 177	125 + 113
....	....	....	....

13.	14.	15.	16.
138 + 188	129 + 190	164 + 163	155 + 111
....	....	....	....

17.	18.	19.	20.
182 + 186	103 + 118	123 + 131	176 + 145
....	....	....	....

21.	22.	23.	24.
152 + 178	156 + 176	187 + 117	110 + 149
....	....	....	....

1. **119+61+11=** ....

2. **38+154+16=** ....

3. **186+69+1+19=** ....

4. **125+53+98=** ....

5. **123+34+61=** ....

6. **68+181+171=** ....

7. **199+13+85+27=** ....

8. **27+46+152=** ....

9. **48+12+74+152=** ....

10. **2+134+113=** ....

11. **50+190+65=** ....

12. **117+113+82=** ....

1. **156+16+44=** ....

2. **42+105+2=** ....

3. **166+91+191=** ....

4. **104+44+47=** ....

5. **48+45+2+188=** ....

6. **62+108+6=** ....

7. **151+71+139=** ....

8. **141+109+172=** ....

9. **8+109+183=** ....

10. **110+140+134=** ....

11. **181+164+63+7=** ....

12. **85+61+101=** ....

1. Jake has 16 apples and Allan has 4 apples. How many apples do Jake and Allan have together?

....

2. 2 balls are in the basket. 13 more balls are put in the basket. How many balls are in the basket now?

....

3. Amy has 26 more oranges than Sandra. Sandra has 32 oranges. How many oranges does Amy have?

....

4. 11 plums were in the basket. More plums were added to the basket. Now there are 48 plums. How many plums were added to the basket?

....

5. 13 peaches were in the basket. 11 are red and the rest are green. How many peaches are green?

....

6. 11 red pears and 45 green pears are in the basket. How many pears are in the basket?

....

7. Some marbles were in the basket. 10 more marbles were added to the basket. Now there are 20 marbles. How many marbles were in the basket before more marbles were added?

....

1. Some oranges were in the basket. 4 more oranges were added to the basket. Now there are 27 oranges. How many oranges were in the basket before more oranges were added?

....

2. Sandra has 17 more marbles than Ellen. Ellen has 5 marbles. How many marbles does Sandra have?

....

3. 20 pears were in the basket. More pears were added to the basket. Now there are 52 pears. How many pears were added to the basket?

....

4. 13 red balls and 33 green balls are in the basket. How many balls are in the basket?

....

5. Brian has 30 peaches and Steven has 37 peaches. How many peaches do Brian and Steven have together?

....

6. 27 plums were in the basket. 22 are red and the rest are green. How many plums are green?

....

7. 19 apples are in the basket. 21 more apples are put in the basket. How many apples are in the basket now?

....

1. Some plums were in the basket. 26 more plums were added to the basket. Now there are 63 plums. How many plums were in the basket before more plums were added?

....

2. Adam has 28 balls and Brian has 27 balls. How many balls do Adam and Brian have together?

....

3. 13 peaches are in the basket. 25 more peaches are put in the basket. How many peaches are in the basket now?

....

4. 81 apples were in the basket. 32 are red and the rest are green. How many apples are green?

....

5. Sharon has 4 more oranges than Marcie. Marcie has 23 oranges. How many oranges does Sharon have?

....

6. 16 pears were in the basket. More pears were added to the basket. Now there are 40 pears. How many pears were added to the basket?

....

7. 9 red marbles and 50 green marbles are in the basket. How many marbles are in the basket?

....

1. 51 oranges were in the basket. 30 are red and the rest are green. How many oranges are green?

2. 20 apples were in the basket. More apples were added to the basket. Now there are 53 apples. How many apples were added to the basket?

3. Jackie has 18 more balls than Marin. Marin has 49 balls. How many balls does Jackie have?

4. 25 red pears and 49 green pears are in the basket. How many pears are in the basket?

5. David has 30 marbles and Brian has 42 marbles. How many marbles do David and Brian have together?

6. 17 plums are in the basket. 2 more plums are put in the basket. How many plums are in the basket now?

7. Some peaches were in the basket. 25 more peaches were added to the basket. Now there are 55 peaches. How many peaches were in the basket before more peaches were added?

SUBTRACTION

1. 35 - 30 = ...

2. 44 - 20 = ...

3. 39 - 23 = ...

4. 31 - 20 = ...

5. 4 - 4 = ...

6. 34 - 3 = ...

7. 29 - 2 = ...

8. 3 - 2 = ...

9. 39 - 4 = ...

10. 24 - 11 = ...

11. 47 - 23 = ...

12. 14 - 3 = ...

13. 11 - 8 = ...

14. 8 - 7 = ...

15. 47 - 20 = ...

16. 18 - 12 = ...

17. 22 - 21 = ...

18. 44 - 40 = ...

19. 43 - 8 = ...

20. 32 - 17 = ...

21. 34 - 16 = ...

22. 50 - 8 = ...

23. 30 - 12 = ...

24. 33 - 32 = ...

1. 24 - 23 = ...

2. 38 - 19 = ...

3. 25 - 4 = ...

4. 27 - 22 = ...

5. 2 - 1 = ...

6. 24 - 11 = ...

7. 20 - 16 = ...

8. 18 - 6 = ...

9. 7 - 5 = ...

10. 17 - 3 = ...

11. 47 - 25 = ...

12. 5 - 5 = ...

13. 38 - 13 = ...

14. 30 - 17 = ...

15. 38 - 2 = ...

16. 40 - 10 = ...

17. 47 - 20 = ...

18. 6 - 3 = ...

19. 20 - 7 = ...

20. 36 - 17 = ...

21. 11 - 5 = ...

22. 5 - 2 = ...

23. 16 - 5 = ...

24. 15 - 12 = ...

1. $23 - 11 =$ ....

2. $37 - 37 =$ ....

3. $52 - 19 =$ ....

4. $64 - 55 =$ ....

5. $87 - 65 =$ ....

6. $29 - 25 =$ ....

7. $92 - 79 =$ ....

8. $12 - 4 =$ ....

9. $93 - 73 =$ ....

10. $85 - 77 =$ ....

11. $36 - 17 =$ ....

12. $52 - 41 =$ ....

13. $76 - 33 =$ ....

14. $38 - 15 =$ ....

15. $31 - 6 =$ ....

16. $9 - 7 =$ ....

17. $35 - 4 =$ ....

18. $95 - 11 =$ ....

19. $20 - 6 =$ ....

20. $89 - 55 =$ ....

21. $44 - 20 =$ ....

22. $28 - 8 =$ ....

23. $70 - 36 =$ ....

24. $94 - 80 =$ ....

1. $\begin{array}{r} 89 \\ -\ 21 \\ \hline \end{array}$ ....

2. $\begin{array}{r} 47 \\ -\ 43 \\ \hline \end{array}$ ....

3. $\begin{array}{r} 20 \\ -\ 6 \\ \hline \end{array}$ ....

4. $\begin{array}{r} 9 \\ -\ 3 \\ \hline \end{array}$ ....

5. $\begin{array}{r} 50 \\ -\ 27 \\ \hline \end{array}$ ....

6. $\begin{array}{r} 16 \\ -\ 12 \\ \hline \end{array}$ ....

7. $\begin{array}{r} 17 \\ -\ 1 \\ \hline \end{array}$ ....

8. $\begin{array}{r} 88 \\ -\ 5 \\ \hline \end{array}$ ....

9. $\begin{array}{r} 7 \\ -\ 2 \\ \hline \end{array}$ ....

10. $\begin{array}{r} 15 \\ -\ 9 \\ \hline \end{array}$ ....

11. $\begin{array}{r} 3 \\ -\ 3 \\ \hline \end{array}$ ....

12. $\begin{array}{r} 64 \\ -\ 12 \\ \hline \end{array}$ ....

13. $\begin{array}{r} 89 \\ -\ 50 \\ \hline \end{array}$ ....

14. $\begin{array}{r} 58 \\ -\ 7 \\ \hline \end{array}$ ....

15. $\begin{array}{r} 98 \\ -\ 10 \\ \hline \end{array}$ ....

16. $\begin{array}{r} 94 \\ -\ 21 \\ \hline \end{array}$ ....

17. $\begin{array}{r} 51 \\ -\ 17 \\ \hline \end{array}$ ....

18. $\begin{array}{r} 20 \\ -\ 12 \\ \hline \end{array}$ ....

19. $\begin{array}{r} 83 \\ -\ 9 \\ \hline \end{array}$ ....

20. $\begin{array}{r} 56 \\ -\ 39 \\ \hline \end{array}$ ....

21. $\begin{array}{r} 52 \\ -\ 51 \\ \hline \end{array}$ ....

22. $\begin{array}{r} 59 \\ -\ 27 \\ \hline \end{array}$ ....

23. $\begin{array}{r} 12 \\ -\ 10 \\ \hline \end{array}$ ....

24. $\begin{array}{r} 55 \\ -\ 37 \\ \hline \end{array}$ ....

1. 11 - 1	2. 32 - 1	3. 90 - 82	4. 66 - 51 ....
5. 23 - 5	6. 108 - 92	7. 115 - 59	8. 51 - 3 ....
9. 63 - 56	10. 55 - 42	11. 91 - 52	12. 49 - 21 ....
13. 48 - 28	14. 86 - 28	15. 130 - 56	16. 104 - 46 ....
17. 86 - 8	18. 147 - 39	19. 34 - 11	20. 117 - 84 ....
21. 21 - 19	22. 119 - 65	23. 60 - 55	24. 40 - 12 ....

1. 73 - 5	2. 89 - 71	3. 14 - 14	4. 86 - 59 ....
5. 137 - 45	6. 21 - 18	7. 37 - 21	8. 66 - 41 ....
9. 63 - 29	10. 26 - 17	11. 26 - 4	12. 55 - 6 ....
13. 25 - 5	14. 34 - 17	15. 129 - 95	16. 90 - 64 ....
17. 60 - 42	18. 42 - 4	19. 42 - 12	20. 71 - 7 ....
21. 77 - 9	22. 86 - 36	23. 107 - 35	24. 19 - 1 ....

1. 115 - 27 = ...

2. 75 - 35 = ...

3. 109 - 72 = ...

4. 192 - 106 = ...

5. 189 - 98 = ...

6. 124 - 100 = ...

7. 155 - 44 = ...

8. 135 - 66 = ...

9. 162 - 135 = ...

10. 165 - 33 = ...

11. 146 - 24 = ...

12. 192 - 12 = ...

13. 77 - 48 = ...

14. 100 - 92 = ...

15. 187 - 82 = ...

16. 163 - 142 = ...

17. 77 - 72 = ...

18. 102 - 31 = ...

19. 97 - 88 = ...

20. 116 - 84 = ...

21. 96 - 31 = ...

22. 185 - 124 = ...

23. 154 - 33 = ...

24. 170 - 92 = ...

1. 118 - 18 = ...

2. 178 - 65 = ...

3. 80 - 20 = ...

4. 98 - 11 = ...

5. 165 - 114 = ...

6. 55 - 51 = ...

7. 139 - 121 = ...

8. 192 - 98 = ...

9. 70 - 11 = ...

10. 53 - 17 = ...

11. 153 - 97 = ...

12. 110 - 37 = ...

13. 121 - 64 = ...

14. 190 - 30 = ...

15. 133 - 14 = ...

16. 154 - 85 = ...

17. 179 - 19 = ...

18. 116 - 108 = ...

19. 154 - 67 = ...

20. 69 - 14 = ...

21. 175 - 37 = ...

22. 58 - 37 = ...

23. 51 - 45 = ...

24. 173 - 12 = ...

1. 38 plums are in the basket. 31 are red and the rest are green. How many plums are green?

....

2. Sharon has 3 fewer peaches than Amy. Amy has 12 peaches. How many peaches does Sharon have?

....

3. Some balls were in the basket. 26 balls were taken from the basket. Now there are 14 balls. How many balls were in the basket before some of the balls were taken?

....

4. Steven has 4 oranges. Jake has 7 oranges. How many more oranges does Jake have than Steven?

....

5. 24 marbles are in the basket. 9 marbles are taken out of the basket. How many marbles are in the basket now?

....

6. 23 apples were in the basket. Some of the apples were removed from the basket. Now there are 4 apples. How many apples were removed from the basket?

....

7. 34 pears are in the basket. 13 pears are taken out of the basket. How many pears are in the basket now?

....

1. Adam has 6 peaches. Billy has 7 peaches. How many more peaches does Billy have than Adam?

....

2. Some balls were in the basket. 24 balls were taken from the basket. Now there are 11 balls. How many balls were in the basket before some of the balls were taken?

....

3. 11 plums were in the basket. Some of the plums were removed from the basket. Now there are 2 plums. How many plums were removed from the basket?

....

4. Jennifer has 4 fewer oranges than Jackie. Jackie has 18 oranges. How many oranges does Jennifer have?

....

5. 22 pears are in the basket. 6 are red and the rest are green. How many pears are green?

....

6. 12 apples are in the basket. 6 apples are taken out of the basket. How many apples are in the basket now?

....

7. Some marbles were in the basket. 16 marbles were taken from the basket. Now there are 22 marbles. How many marbles were in the basket before some of the marbles were taken?

....

1. Janet has 7 fewer marbles than Michele. Michele has 23 marbles. How many marbles does Janet have?

....

2. 5 pears are in the basket. 3 pears are taken out of the basket. How many pears are in the basket now?

....

3. 4 oranges are in the basket. 4 are red and the rest are green. How many oranges are green?

....

4. Some plums were in the basket. 4 plums were taken from the basket. Now there are 13 plums. How many plums were in the basket before some of the plums were taken?

....

5. 11 peaches were in the basket. Some of the peaches were removed from the basket. Now there are 7 peaches. How many peaches were removed from the basket?

....

6. Steven has 15 apples. David has 18 apples. How many more apples does David have than Steven?

....

7. 22 balls are in the basket. 4 are red and the rest are green. How many balls are green?

....

1. Amy has 7 fewer oranges than Sandra. Sandra has 38 oranges. How many oranges does Amy have?

....

2. Some peaches were in the basket. 26 peaches were taken from the basket. Now there are 3 peaches. How many peaches were in the basket before some of the peaches were taken?

....

3. Allan has 2 marbles. Donald has 39 marbles. How many more marbles does Donald have than Allan?

....

4. 27 plums are in the basket. 10 are red and the rest are green. How many plums are green?

....

5. 22 apples were in the basket. Some of the apples were removed from the basket. Now there are 0 apples. How many apples were removed from the basket?

....

6. 32 pears are in the basket. 32 pears are taken out of the basket. How many pears are in the basket now?

....

7. Brian has 10 balls. Paul has 10 balls. How many more balls does Paul have than Brian?

....

MULTIPLICATION

1. 14×7 = ...

2. 14×3 = ...

3. 11×6 = ...

4. 5×7 = ...

5. 12×9 = ...

6. 4×3 = ...

7. 9×2 = ...

8. 13×9 = ...

9. 5×1 = ...

10. 12×2 = ...

11. 17×3 = ...

12. 19×5 = ...

13. 3×9 = ...

14. 4×5 = ...

15. 5×4 = ...

16. 1×10 = ...

17. 7×10 = ...

18. 8×5 = ...

19. 18×7 = ...

20. 18×4 = ...

21. 10×9 = ...

22. 14×5 = ...

23. 8×1 = ...

24. 18×6 = ...

1. 11×4 = ...

2. 22×3 = ...

3. 98×1 = ...

4. 20×4 = ...

5. 33×2 = ...

6. 88×1 = ...

7. 10×3 = ...

8. 10×2 = ...

9. 42×2 = ...

10. 10×5 = ...

11. 13×1 = ...

12. 23×2 = ...

13. 12×3 = ...

14. 44×2 = ...

15. 21×4 = ...

16. 40×2 = ...

17. 22×4 = ...

18. 23×3 = ...

19. 12×2 = ...

20. 30×3 = ...

21. 30×2 = ...

22. 13×2 = ...

23. 24×2 = ...

24. 11×5 = ...

1. 10×4 = ...

2. 13×2 = ...

3. 11×5 = ...

4. 23×3 = ...

5. 21×4 = ...

6. 11×4 = ...

7. 32×3 = ...

8. 22×4 = ...

9. 12×4 = ...

10. 11×2 = ...

11. 30×3 = ...

12. 22×3 = ...

13. 40×2 = ...

14. 33×3 = ...

15. 10×5 = ...

16. 42×2 = ...

17. 10×3 = ...

18. 11×3 = ...

19. 25×1 = ...

20. 43×2 = ...

21. 12×2 = ...

22. 44×2 = ...

23. 41×1 = ...

24. 30×2 = ...

1. $22 \times 4 =$ ....

2. $31 \times 3 =$ ....

3. $11 \times 4 =$ ....

4. $22 \times 3 =$ ....

5. $32 \times 2 =$ ....

6. $20 \times 4 =$ ....

7. $13 \times 3 =$ ....

8. $21 \times 2 =$ ....

9. $12 \times 3 =$ ....

10. $10 \times 4 =$ ....

11. $30 \times 3 =$ ....

12. $23 \times 3 =$ ....

13. $10 \times 3 =$ ....

14. $43 \times 1 =$ ....

15. $24 \times 2 =$ ....

16. $12 \times 4 =$ ....

17. $33 \times 2 =$ ....

18. $94 \times 1 =$ ....

19. $21 \times 4 =$ ....

20. $42 \times 2 =$ ....

21. $76 \times 1 =$ ....

22. $41 \times 2 =$ ....

23. $10 \times 5 =$ ....

24. $34 \times 2 =$ ....

1. 132 × 2 = ...

2. 101 × 4 = ...

3. 221 × 3 = ...

4. 331 × 3 = ...

5. 201 × 3 = ...

6. 320 × 3 = ...

7. 111 × 3 = ...

8. 111 × 4 = ...

9. 211 × 4 = ...

10. 202 × 4 = ...

11. 220 × 4 = ...

12. 720 × 1 = ...

13. 103 × 2 = ...

14. 233 × 3 = ...

15. 200 × 3 = ...

16. 122 × 4 = ...

17. 221 × 4 = ...

18. 222 × 3 = ...

19. 120 × 4 = ...

20. 402 × 2 = ...

21. 210 × 3 = ...

22. 113 × 2 = ...

23. 210 × 4 = ...

24. 201 × 4 = ...

1. $222 \times 4 =$ ....

2. $202 \times 2 =$ ....

3. $110 \times 3 =$ ....

4. $222 \times 3 =$ ....

5. $100 \times 4 =$ ....

6. $232 \times 2 =$ ....

7. $332 \times 2 =$ ....

8. $220 \times 3 =$ ....

9. $101 \times 5 =$ ....

10. $330 \times 2 =$ ....

11. $213 \times 3 =$ ....

12. $201 \times 4 =$ ....

13. $231 \times 3 =$ ....

14. $270 \times 1 =$ ....

15. $221 \times 4 =$ ....

16. $333 \times 3 =$ ....

17. $330 \times 3 =$ ....

18. $101 \times 4 =$ ....

19. $440 \times 2 =$ ....

20. $200 \times 4 =$ ....

21. $334 \times 2 =$ ....

22. $111 \times 5 =$ ....

23. $322 \times 1 =$ ....

24. $210 \times 4 =$ ....

1. 19 × 10 = ...	2. 19 × 6 = ...	3. 20 × 16 = ...	4. 19 × 4 = ...
5. 20 × 8 = ...	6. 10 × 6 = ...	7. 15 × 12 = ...	8. 17 × 15 = ...
9. 15 × 13 = ...	10. 15 × 3 = ...	11. 10 × 5 = ...	12. 10 × 10 = ...
13. 20 × 4 = ...	14. 15 × 5 = ...	15. 17 × 5 = ...	16. 19 × 13 = ...
17. 12 × 17 = ...	18. 16 × 16 = ...	19. 13 × 20 = ...	20. 18 × 6 = ...
21. 16 × 9 = ...	22. 14 × 15 = ...	23. 13 × 11 = ...	24. 14 × 9 = ...

1. $18 \times 7 =$ ....

2. $14 \times 14 =$ ....

3. $17 \times 8 =$ ....

4. $11 \times 11 =$ ....

5. $17 \times 4 =$ ....

6. $20 \times 8 =$ ....

7. $11 \times 17 =$ ....

8. $11 \times 7 =$ ....

9. $14 \times 18 =$ ....

10. $15 \times 16 =$ ....

11. $12 \times 3 =$ ....

12. $16 \times 17 =$ ....

13. $18 \times 12 =$ ....

14. $14 \times 2 =$ ....

15. $15 \times 15 =$ ....

16. $14 \times 7 =$ ....

17. $20 \times 11 =$ ....

18. $12 \times 7 =$ ....

19. $20 \times 18 =$ ....

20. $10 \times 11 =$ ....

21. $17 \times 11 =$ ....

22. $12 \times 16 =$ ....

23. $13 \times 3 =$ ....

24. $14 \times 13 =$ ....

1. David can cycle 17 miles per hour. How far can David cycle in 8 hours?

....

2. David swims 15 laps every day. How many laps will David swim in 5 days?

....

3. If there are 8 oranges in each box and there are 14 boxes, how many oranges are there in total?

....

4. Marcie has 14 times more peaches than Ellen. Ellen has 18 peaches. How many peaches does Marcie have?

....

5. Sharon's garden has 12 rows of pumpkins. Each row has 18 pumpkins. How many pumpkins does Sharon have in all?

....

6. Jennifer's garden has 10 rows of pumpkins. Each row has 10 pumpkins. How many pumpkins does Jennifer have in all?

....

7. Allan swims 19 laps every day. How many laps will Allan swim in 18 days?

....

1. Sandra's garden has 7 rows of pumpkins. Each row has 3 pumpkins. How many pumpkins does Sandra have in all?

....

2. Jackie swims 19 laps every day. How many laps will Jackie swim in 20 days?

....

3. Amy has 3 times more plums than Sandra. Sandra has 1 plum. How many plums does Amy have?

....

4. Donald can cycle 10 miles per hour. How far can Donald cycle in 19 hours?

....

5. If there are 20 marbles in each box and there are 3 boxes, how many marbles are there in total?

....

6. Billy has 9 times more pears than Marcie. Marcie has 5 pears. How many pears does Billy have?

....

7. If there are 17 balls in each box and there are 3 boxes, how many balls are there in total?

....

1. Donald has 6 times more plums than Michele. Michele has 13 plums. How many plums does Donald have?

....

2. If there are 10 marbles in each box and there are 15 boxes, how many marbles are there in total?

....

3. Sandra swims 17 laps every day. How many laps will Sandra swim in 12 days?

....

4. Allan can cycle 2 miles per hour. How far can Allan cycle in 11 hours?

....

5. Marin's garden has 8 rows of pumpkins. Each row has 5 pumpkins. How many pumpkins does Marin have in all?

....

6. David has 11 times more peaches than Allan. Allan has 5 peaches. How many peaches does David have?

....

7. Amy swims 12 laps every day. How many laps will Amy swim in 11 days?

....

1. Janet has 8 times more peaches than Donald. Donald has 16 peaches. How many peaches does Janet have?

....

2. Jake can cycle 12 miles per hour. How far can Jake cycle in 17 hours?

....

3. Jennifer's garden has 16 rows of pumpkins. Each row has 11 pumpkins. How many pumpkins does Jennifer have in all?

....

4. If there are 19 balls in each box and there are 15 boxes, how many balls are there in total?

....

5. Adam swims 1 laps every day. How many laps will Adam swim in 3 days?

....

6. Paul can cycle 9 miles per hour. How far can Paul cycle in 15 hours?

....

7. Michele's garden has 14 rows of pumpkins. Each row has 14 pumpkins. How many pumpkins does Michele have in all?

....

DIVISION

1. **14 ÷ 7 =** ....

2. **12 ÷ 6 =** ....

3. **6 ÷ 3 =** ....

4. **56 ÷ 8 =** ....

5. **49 ÷ 7 =** ....

6. **24 ÷ 6 =** ....

7. **5 ÷ 1 =** ....

8. **21 ÷ 7 =** ....

9. **30 ÷ 6 =** ....

10. **6 ÷ 6 =** ....

11. **36 ÷ 9 =** ....

12. **40 ÷ 5 =** ....

13. **16 ÷ 2 =** ....

14. **24 ÷ 3 =** ....

15. **24 ÷ 4 =** ....

16. **1 ÷ 1 =** ....

17. **35 ÷ 7 =** ....

18. **10 ÷ 2 =** ....

19. **56 ÷ 7 =** ....

20. **14 ÷ 2 =** ....

21. **2 ÷ 1 =** ....

22. **42 ÷ 7 =** ....

23. **4 ÷ 4 =** ....

24. **20 ÷ 5 =** ....

1. **72 ÷ 9 =** ....

2. **48 ÷ 6 =** ....

3. **5 ÷ 5 =** ....

4. **30 ÷ 5 =** ....

5. **27 ÷ 3 =** ....

6. **40 ÷ 5 =** ....

7. **28 ÷ 4 =** ....

8. **32 ÷ 8 =** ....

9. **8 ÷ 8 =** ....

10. **6 ÷ 3 =** ....

11. **42 ÷ 7 =** ....

12. **12 ÷ 3 =** ....

13. **12 ÷ 4 =** ....

14. **15 ÷ 5 =** ....

15. **35 ÷ 7 =** ....

16. **21 ÷ 7 =** ....

17. **24 ÷ 6 =** ....

18. **25 ÷ 5 =** ....

19. **20 ÷ 4 =** ....

20. **36 ÷ 9 =** ....

21. **42 ÷ 6 =** ....

22. **1 ÷ 1 =** ....

23. **10 ÷ 5 =** ....

24. **40 ÷ 8 =** ....

1. **12 ÷ 3 =** ....
2. **9 ÷ 9 =** ....
3. **18 ÷ 3 =** ....
4. **24 ÷ 3 =** ....
5. **1 ÷ 1 =** ....
6. **30 ÷ 6 =** ....
7. **24 ÷ 8 =** ....
8. **16 ÷ 4 =** ....
9. **49 ÷ 7 =** ....
10. **2 ÷ 2 =** ....
11. **16 ÷ 8 =** ....
12. **14 ÷ 2 =** ....
13. **56 ÷ 8 =** ....
14. **10 ÷ 2 =** ....
15. **6 ÷ 3 =** ....
16. **63 ÷ 9 =** ....
17. **4 ÷ 2 =** ....
18. **35 ÷ 7 =** ....
19. **40 ÷ 8 =** ....
20. **14 ÷ 7 =** ....
21. **40 ÷ 5 =** ....
22. **20 ÷ 4 =** ....
23. **25 ÷ 5 =** ....
24. **32 ÷ 8 =** ....

1. **18 ÷ 2 =** ....
2. **42 ÷ 6 =** ....
3. **16 ÷ 8 =** ....
4. **40 ÷ 4 =** ....
5. **48 ÷ 8 =** ....
6. **13 ÷ 1 =** ....
7. **112 ÷ 8 =** ....
8. **6 ÷ 3 =** ....
9. **32 ÷ 8 =** ....
10. **72 ÷ 9 =** ....
11. **30 ÷ 3 =** ....
12. **25 ÷ 5 =** ....
13. **45 ÷ 5 =** ....
14. **7 ÷ 1 =** ....
15. **35 ÷ 5 =** ....
16. **54 ÷ 6 =** ....
17. **70 ÷ 7 =** ....
18. **64 ÷ 8 =** ....
19. **14 ÷ 2 =** ....
20. **12 ÷ 6 =** ....
21. **9 ÷ 3 =** ....
22. **108 ÷ 9 =** ....
23. **24 ÷ 2 =** ....
24. **16 ÷ 4 =** ....

1. **69 ÷ 3 =** ....

2. **147 ÷ 7 =** ....

3. **38 ÷ 2 =** ....

4. **114 ÷ 6 =** ....

5. **108 ÷ 4 =** ....

6. **112 ÷ 8 =** ....

7. **175 ÷ 7 =** ....

8. **76 ÷ 4 =** ....

9. **14 ÷ 7 =** ....

10. **65 ÷ 5 =** ....

11. **52 ÷ 4 =** ....

12. **56 ÷ 7 =** ....

13. **85 ÷ 5 =** ....

14. **116 ÷ 4 =** ....

15. **144 ÷ 9 =** ....

16. **75 ÷ 5 =** ....

17. **36 ÷ 6 =** ....

18. **48 ÷ 4 =** ....

19. **30 ÷ 3 =** ....

20. **207 ÷ 9 =** ....

21. **87 ÷ 3 =** ....

22. **63 ÷ 9 =** ....

23. **14 ÷ 2 =** ....

24. **196 ÷ 7 =** ....

1. **116 ÷ 4 =** ....

2. **63 ÷ 9 =** ....

3. **66 ÷ 6 =** ....

4. **20 ÷ 4 =** ....

5. **6 ÷ 3 =** ....

6. **174 ÷ 6 =** ....

7. **189 ÷ 7 =** ....

8. **36 ÷ 6 =** ....

9. **162 ÷ 6 =** ....

10. **18 ÷ 3 =** ....

11. **18 ÷ 1 =** ....

12. **208 ÷ 8 =** ....

13. **21 ÷ 1 =** ....

14. **26 ÷ 2 =** ....

15. **12 ÷ 3 =** ....

16. **72 ÷ 3 =** ....

17. **92 ÷ 4 =** ....

18. **15 ÷ 5 =** ....

19. **50 ÷ 5 =** ....

20. **161 ÷ 7 =** ....

21. **30 ÷ 6 =** ....

22. **54 ÷ 2 =** ....

23. **196 ÷ 7 =** ....

24. **20 ÷ 5 =** ....

1. **30 ÷ 2 =** ....

2. **135 ÷ 5 =** ....

3. **116 ÷ 4 =** ....

4. **4 ÷ 1 =** ....

5. **6 ÷ 3 =** ....

6. **48 ÷ 4 =** ....

7. **26 ÷ 2 =** ....

8. **96 ÷ 8 =** ....

9. **133 ÷ 7 =** ....

10. **50 ÷ 2 =** ....

11. **108 ÷ 6 =** ....

12. **9 ÷ 3 =** ....

13. **4 ÷ 2 =** ....

14. **64 ÷ 4 =** ....

15. **84 ÷ 6 =** ....

16. **38 ÷ 2 =** ....

17. **84 ÷ 7 =** ....

18. **10 ÷ 5 =** ....

19. **24 ÷ 4 =** ....

20. **3 ÷ 1 =** ....

21. **35 ÷ 7 =** ....

22. **96 ÷ 6 =** ....

23. **108 ÷ 9 =** ....

24. **56 ÷ 8 =** ....

1. **12 ÷ 4 =** ....
2. **15 ÷ 1 =** ....
3. **168 ÷ 8 =** ....
4. **46 ÷ 2 =** ....
5. **12 ÷ 2 =** ....
6. **104 ÷ 4 =** ....
7. **189 ÷ 7 =** ....
8. **52 ÷ 4 =** ....
9. **75 ÷ 5 =** ....
10. **16 ÷ 4 =** ....
11. **4 ÷ 4 =** ....
12. **12 ÷ 3 =** ....
13. **168 ÷ 7 =** ....
14. **42 ÷ 3 =** ....
15. **60 ÷ 6 =** ....
16. **162 ÷ 6 =** ....
17. **48 ÷ 8 =** ....
18. **39 ÷ 3 =** ....
19. **2 ÷ 1 =** ....
20. **144 ÷ 8 =** ....
21. **64 ÷ 4 =** ....
22. **22 ÷ 2 =** ....
23. **20 ÷ 2 =** ....
24. **14 ÷ 1 =** ....

1. A box of pears weighs 70 pounds. If one pears weighs 7 pounds, how many pears are there in the box?

....

2. You have 30 marbles and want to share them equally with 10 people. How many marbles would each person get?

....

3. Jackie made 30 cookies for a bake sale. She put the cookies in bags, with 6 cookies in each bag. How many bags did she have for the bake sale?

....

4. Jackie ordered 7 pizzas. The bill for the pizzas came to $70. What was the cost of each pizza?

....

5. How many 3 cm pieces of rope can you cut from a rope that is 12 cm long?

....

6. Jake is reading a book with 28 pages. If Jake wants to read the same number of pages every day, how many pages would Jake have to read each day to finish in 4 days?

....

7. David ordered 5 pizzas. The bill for the pizzas came to $5. What was the cost of each pizza?

....

1. How many 7 cm pieces of rope can you cut from a rope that is 14 cm long?

....

2. You have 20 marbles and want to share them equally with 5 people. How many marbles would each person get?

....

3. Adam is reading a book with 12 pages. If Adam wants to read the same number of pages every day, how many pages would Adam have to read each day to finish in 3 days?

....

4. Billy ordered 2 pizzas. The bill for the pizzas came to $4. What was the cost of each pizza?

....

5. Janet made 7 cookies for a bake sale. She put the cookies in bags, with 7 cookies in each bag. How many bags did she have for the bake sale?

....

6. A box of balls weighs 63 pounds. If one balls weighs 9 pounds, how many balls are there in the box?

....

7. How many 4 cm pieces of rope can you cut from a rope that is 4 cm long?

....

1. You have 32 balls and want to share them equally with 8 people. How many balls would each person get?

....

2. Sandra ordered 1 pizzas. The bill for the pizzas came to $2. What was the cost of each pizza?

....

3. Janet made 12 cookies for a bake sale. She put the cookies in bags, with 2 cookies in each bag. How many bags did she have for the bake sale?

....

4. Donald is reading a book with 5 pages. If Donald wants to read the same number of pages every day, how many pages would Donald have to read each day to finish in 5 days?

....

5. How many 3 cm pieces of rope can you cut from a rope that is 18 cm long?

....

6. A box of apples weighs 4 pounds. If one apples weighs 4 pounds, how many apples are there in the box?

....

7. You have 56 marbles and want to share them equally with 8 people. How many marbles would each person get?

....

1. Marin made 20 cookies for a bake sale. She put the cookies in bags, with 2 cookies in each bag. How many bags did she have for the bake sale?

....

2. You have 12 apples and want to share them equally with 2 people. How many apples would each person get?

....

3. How many 8 cm pieces of rope can you cut from a rope that is 24 cm long?

....

4. David is reading a book with 21 pages. If David wants to read the same number of pages every day, how many pages would David have to read each day to finish in 3 days?

....

5. Jackie ordered 9 pizzas. The bill for the pizzas came to $36. What was the cost of each pizza?

....

6. A box of balls weighs 9 pounds. If one balls weighs 9 pounds, how many balls are there in the box?

....

7. You have 21 marbles and want to share them equally with 3 people. How many marbles would each person get?

....

Made in the USA
Las Vegas, NV
18 March 2023